S0-ANP-712

Robin Becker
Helena Minton
Marilyn Zuckerman

Personal Effects

Library of Congress Catalogue Card Number
75-46406
ISBN: 0-914086-15-4
Printed in the United States of America

Book design by Don Kline
Cover painting by Gwen Fabricant

The publication of this book was supported by a grant from the National Endowment for the Arts, Washington, D.C., and assisted by the Massachusetts Council for the Arts and Humanities.

Special thanks to New Hampshire Composition, Inc.

Alice James Books are published by Alice James Poetry Cooperative, Inc.

ALICE JAMES BOOKS
Cambridge, Massachusetts

Personal Effects

Thanks are due to the editors of the following publications in which versions of these poems originally appeared:

Robin Becker

AMAZON POETRY: AN ANTHOLOGY, 1975 (Out & Out Books), APHRA, Aspect and The Second Wave.

Many thanks to Cummington Community of the Arts where some of these poems were written.

Helena Minton

Aspect, Chomo Uri, The Chowder Review, Dark Horse, Epos, Four Zoas, The Greenfield Review, INTRO #6, Mirror Northwest, New America: A Review, Panache, The Seneca Review, Spectrum, Stonecloud, The Woman's Journal and *Flowering After Frost: An Anthology of New England Poets*, 1975 (The Branden Press).

And special thanks to David De'Innocentis.

Marilyn Zuckerman

Connections, New York Quarterly, Sunbury I, and an anthology, *The City of Angels*, 1976 (Los Angeles Review).

Robin Becker

Discretion

Contents

Robin Becker

Discretion

The Letter

Every dispatch courses its way to consciousness
like something outrageous done for love.

A yellow office memo, sealed and moving
from hand to hand always arrives, is always arriving...
In Cullera, in the flux of summer things,
time passed like blue sailboats on green water;
reason and plan forgotten.

At noon, a single mind or ravaged place
contends with small histories; the habit
of a restaurant, books arranged in a certain order
to catalogue the arbitrary.
What unrecognized sport!
At once examiner to interpret the data and
examinee to withstand the emotion.

Here, take it. It is for you. If there is no shadow
of a doubt, beyond a shadow of a doubt,
then the letter is yours and has been searching for you.
Do you remember when you mailed it from your dream,
the one in which your arm had been severed,
and you stared at the dried blood on the bone, wondering
how it could be your own
arm, blood, bone?

November

Of the cold of November this year,
Last, others, little breakthroughs like
Shadows on the wall; twice the number of dried flowers,
Couplets of coffee mugs,
Ashtrays reflecting dimensionless doubles.

Across the evening, a voice
Reflects a memory of that voice.
Someone prepares dinner.
A figure passes behind the crisscross
Of a window.

It's winter. We are walking
From my car parked in some alley.
Back-to-back a fantasy—
That going away
That coming together.

This city hurries to the shortest day of the year.
We are dogs chasing our tails,
Eating ourselves in the cold,
Eating our lunches and dinners,
Eating books and reflections,
Eating the past, the visions of the past,
Eating the visions of the visions of the past,
Eating the night, eating without end.

A Woman Leaving A Woman

You are setting out from Cambridge
with an old debt
in a brown hat
running like a collapsible umbrella,
a laundry bag of detail slung across your back.

For weeks, you have told me of stirring dramas.
One is strewn with landmarks; years are there.
(He is a man; his life makes sense;
he has come for you; you will go.)

Departure is a striped shirt, figuring
into the crowd, a foreigner checking the maps.

You're carrying it off, sister,
like an exercise in fidelity,
like Indian summer in mid-November.
In a few years we'll laugh, shrug,
but for now,

you're carrying it off;
quickly breakfasting with someone else's children
as if a crime had been committed, and you
were the only witness.

If you return,
if you do not,
if you discover one voice
to satisfy your ache for a chorus,
some morning you will find sandpaper scratching
your memory.

Swallow. It will be your own throat.
Touch yourself. It will be your own hands
serving your own body.
Finally, it will make you glad.

Afterthought

I

Houseguesting
through the summer,
I made friends with the cat.
In July, she frequented
my room on her way
to the trees.

This fall, she turns huntress,
snagging feathers in her whiskers,
smug, not interested in dinner.

When I go out to look for work,
I look for passion.
It is a calling I ignore
as long as I can,
but it's bound to get me.

II

Old passions and old lovers
ought not to return.
Decoys on the water,
they lure what they cannot love,
trap or kill from a distance,
return nothing.

Two old politicians
sit up late,
lucid, reminiscent.
We seek each other
to embrace, to be solvent and
answerable.
In the embrace, we are not solvent,
but feel so.

The Airport

In flying, there is an even exchange.
For every pair of lovers who part
at the gate, there is a plane
conveying one to another.

I must leave you at Logan so early,
that we arise in the dark to dress.

The Airport is a hometown
between two cities.
It welcomes you, baggage and nerve,
to an extended family
and pushes you on.

You trot down the bubble
and into the plane. You wave goodbye.
I leave you, stalled in a jet.
And in Chicago, someone is waking
to meet you at O'Hare.

For Bubbie Growing Older

These days, your brittle fingers can barely
lace your corset or slice the apples for strudel.
You curse the God of Abraham and Isaac
Who let this happen.   Age.
Age is the long thud on the door.
Age is stiffness in the morning,
an afternoon's painful walk to the market,
where you cannot reach
the only grapefruit worth buying.

Age is the memory
confusing countries
you lived in, muddling
the perfect stories
I grew up on
and now recite
back to you.

Bubbie, when I was a kid,
we stood in the ocean
at Atlantic City.
You were my lifeguard,
my big Bubbie,
throwing me into the waves,
that I should know
how to steer my body
through the undertow.

Big Business

What is business, that we
must take such care with it?
The careful setting out of clothes
the night before, always
more than a grain of salt.

When I was a little girl, my mother
said, *Robbie, you better get to bed*
because you have a big day tomorrow.
That meant business, the dentist
after school, or a violin lesson.

The dog went out to do his business.
He was quick about it.

Mother sat downstairs and worried
about daddy's business.
How's business?

Today, I can't get down to it.
I am rubbing two sticks of business
together, anxious,
waiting for something
to fly out.

Money

By now, everyone has
plans for the evening.
I don't care.
I have an eight-hour day
under my belt.

Daddy, I'll make so much money
we can go back to prep school,
conjugate all the irregular
Latin verbs—*sum . . . es . . . est . . .*

I'll learn the value of a dollar!
There's still time!

Mother, I'll support us
in the style to which we have
become acclimated. We'll go to Florida
for the winter, get a condominium,
take Bubbie.

Bubbie, I'll make us
honest women; all that greed
and no way to spend it.
We'll shop for the jewels the Russian
princesses wore, when you were the tailor's
daughter riding in the oxcart; riding
to deliver the wedding dresses
and the party dresses.

And Bubbie, I've always wanted to ask—
Did you wish, did you ever wish,
that you were not Jewish,
and not living in the shtetl?

For Tess

My cousin is dying   one
by one her organs blink out
She never had a chance
It's downhill from here

Tess my brilliant cousin
laughed at the Seders
came home from New York
to please the family
taught English to policemen
nursed a sick aunt on the East Side
At 29 you have come home to die

Today they tell me not to call you
that your parents will cry
that you've lost your voice
your legs   your sight   your job

As the other girl cousins got married
had kids   you got a degree
and moved to the City
They say you *knew*
never got married because of *the illness*

In junior high I bragged to my friends
my cousin walking MacDougal Street
with her bookbag   Tess
who chose to spend the summer at Columbia
while the uncles and cousins packed
for Atlantic City   I remember hearing   then
that your eyes were going bad

Tess   what did you know?
extra sessions for the cops
Old English translations stacked by your bed
weekly trips through the park with old Aunt Sarah.

The Seizure

When the seizures struck,
like a flashflood,
thrust you
stormcenter, I
screamed in your ears,
clapped my hands that you might listen;
deafened, you stared into the volley.

Behind the curtained air,
syllables stuck together;
all your limbs fell off;
bombarded, the torso surrendered.

I clawed at the straitjacket silence,
afraid that I would catch it
and fall from the day
as I might fall from my bicycle;
afraid of the metals
they hooked in your hair
like hoses to pull the empty places
out of you and onto the paper.

Stowaway, you eluded them.
That passage knew you,
spelt your name, like a flashback,
lifted you day after day
out of the afternoon,
unremembering.

The Plan

for Janie

It is April, and this plan is almost over.
We lunch at The Union Oyster House
to talk about the summer;
part-time jobs in the city,
the pros and cons of travelling.

Leaving places gets harder. Admit it.
After the interview, the editor said,
A great set of boobs. You walked
around New York, flipping through
your catalogue of friends to visit.

On the last day of camp, when we were ten,
I cried, riding the bus back home,
like growing up. You read, told me
I was a *baby*. Janie, I could not be like you.

It's April, and the rest of the world
is into something. Like the machines
into which you slide a quarter,
receive two dimes and a nickel,
every plan emits its divisions.
It is the season of change and reduction.

Only you refuse, still believe
in the big events, between times.

Policy

I'm calling because I'm concerned about you. No matter what you do in life, it's important to know your future is secure. I'm talking about Life Insurance.

I appreciate your concern. It's
just that I don't believe
in Life Insurance.

Yes, I want a guarantee,
to guarantee me a full tank of gas,
friends who will not move
to New York or California,
a lover with a life of her own,
and her mind set on me.

Driving through this mountain pass,
thick with fog and mist,
I could so easily crash my car,
never live to tell what this life
of risk taking is all about.

Weekend

I

I am learning to secure myself
to rocks and trees
Naked in a field
steady as the mountains
I let go my lay-away plan
and listen

II

Leaving you more often
than I thought I would
I'm heading west
on the Mass Pike
out of Boston

III

Last night I dreamt of swans
in tandem
Our talk of possession
the danger of coupling
nourished my sleep

IV

And yet it is so good
to be alone here
Tenure in the country
a workplace in the orchard
I am a misplaced citydweller
breathing honeysuckle
and the smell of you
still on my fingers

Waterclock

for Gwen

I

In the tangle of treetrunks
and overgrowth   a stone pool
built for the cautious swimmer

Beneath the pool   water gushes
from stone no hand has fashioned
battering the defiant   churning

We lie on the rocks   waterworn
glistening   above a landscape
of foam and temptation

The roar of the white
water   crashes
in our brains

II

I think of the Moors
triumphant   finding
that place in the mountains

Building streets and squares
to hear water
from every marketplace

I remember my own relief
driving south with Sylvie
a narrow road dusty with chickens

From Madrid to Valencia
The Spanish National Highway
past Elche and up   up

Into the mountains
scrub vegetation
cactus

We finally came to *Granada*
rain forest   waterfall
Alhambra   Alhambra

Bubbie 1975

I see you your red and
white striped beachdress
stooped walking

slower than
last summer Bubbie
I see your faded beachdress
blue thongs straw bag

Can I have a sailor's hat?
Will we buy macaroons?
Is it time to go to the beach?
You dressed us for dinner
and the long evenings
on the boardwalk

This summer I have questions
no one can answer
Flying higher on the ferris wheel
lights tinkling surprises
up up in Atlantic City

Bubbie I've been away
in the woods and I remember
how you used to say
Mein kind, you have your head
in the trees

Four Days of Rain

The sky cracks once and closes.
In the garden, a tiny figure
looks up and hurries away.
The dogs rush in. Someone
runs to shut the windows.
The afternoon goes black.

Dampness like disease is spreading.
For days my socks hang sodden.
Passing from room to room,
I make mental notes—
the roof leaks in the bathroom,
a floorboard is coming up.
I touch a pantsleg; still wet.

Novitiates,
we move about the countryside,
our dark habits
flapping in the wind.

For once all weather reports agree;
no end in sight. *Rain*
I steal an hour out-of-doors
with the mosquitoes to examine
the ground swells, my boots
sinking and sliding in the mud.

I am out in this weather
to recover my faith, to find
a survivor, or a reason
to dance.

Dreams of Smoke and Fire

for Maia

I

A woman dreams of Poland
and wakes up screaming.
You tell me
This is not unusual.
Long after the war
people had nightmares
of scaffolds and torture.

You tell me
In America, you do not know
the meaning of friendship.
We spoke to our friends daily.
Sudden disappearances
were natural.

Today I ate strawberries
with your daughter.
We laughed and talked
of horses, bicycles, camp.
She is an American.

You were brave.
History saw to that.
Now you are surprised
when I need you to tell me
stories of survival,
stories of death.

But we joke
and you make up a story—
I will be rich and
live by the water.

II

In Paris, you took an apartment
with the little girl.
After years of Catholicism,
there are museums, pastry shops
in St. Michel,
for breakfast, long loaves
of bread from the baker.
Two slices of ham were enough.
You were happy.

In the photograph album,
a little girl is dressed like a princess.
The little Princess wears a white dress
and white gloves.

Today she is a teenager
with good sense,
an engineer's cap
and blue jeans.

III

It is Warsaw.
This could have been my life,
but it is not;
it is yours.

I have called you
Survivor. Courageous.
Words like
Ghetto Dachau
sound in dreams of smoke and fire.

Warsaw—
You tell me stories.
I crave them.
Warsaw—
gutted in fires I cannot imagine.

In a photograph,
a child is dressed like a princess
in the Gardens of Warsaw.

In America,
she wears an engineer's cap.
She is happy.

You do not understand.
The Occupation keeps me awake.

The Landing

I

Intimate after weeks of close quarters,
warnings of scurvy, beer running low,
they watched the land mass
grow like another ocean
into sand cliffs one hundred feet high.

That morning, no tavern, no public house
received the ship; etched
on a piece of whalebone, it rocked;
scrimshaw designs of shrouds and stays
thrown into calmer waters.

Five year old Resolved White gave his hand
to his mother and followed her out of the hold.
Oceanus Brewster, born aboard ship,
had not even opened his eyes, when his father
scanned the shore, blinking back mountains
of twilled cotton and flax.

Leyden had made cloth merchants of them;
a handful of English, ill-at-ease
in the lowlands of Holland, turned
adventurers, soldiers of the kingdom
of Christ, with a Church to found,
contracts to draw.

II

From the shoaling
waters, pebbles
foam with the breakers,
slosh and splinter
over the sandbars
to the foreshore.

If you follow
one grain as it
oscillates
in the swash,
focus on
the small matter
of a single
reversal,
you will lose
the grain.

Somewhere, in the back-
wash, in a moment
of tumult, it
returns, over
the bars and
back, over the
bars and back.

III

Dorothy Bradford waited for her husband.
Twice he had waved goodbye; twice
he had returned in the tiny shallop;
he brought tales of adventure at sea,
and once, someone else's harvest,
several sacks of corn.

Six weeks anchored in Provincetown,
she stared at the coast and waited.
Bands of men sailed off to beach on the bars.
Each week brought them closer to winter.

Sealed into the sands of Provincetown's dunes
was her complaint—
that a small child was left behind.
William had insisted. Too good to protest,
too Christian to argue,
she helped the other women with their children,
while the smell of fish—
dead, fresh, decaying—
settled in her nostrils.

IV

They dug at the small mound, pushing
the sand back like dogs
to uncover a bone. Bowls, trays, beads,
a knife, the parcels the dead take
into the ground, they drew and
stuffed into satchels.

They did not uncover
the bones of a sea fowl
but the head and limbs of an Indian child
wrapped in string and bracelets.

V

Dorothy Bradford's Dream

The forest cover gives way. Pitch pines
tumble down the soft shoulders, and beach grass
blows like hair down the broad backs of the dunes.
Two million tons of flashflooding sand
race into the harbor, seal the ship.
The sand advances. I am a bit of sleeve
on a masthead, falling fast.

VI

Sometime during the third expedition,
Dorothy prayed and slipped into the water.
When her husband
returned,
the shipmates tried to explain it.
It was an accident. They were so sorry.
She slipped. No one knew.

Helena Minton

After Curfew

In Memory of my Grandmother,
Alice Cronan Minton
1888-1976

Contents

Helena Minton

After Curfew

September

Zucchini fattens inedible.
Beans and apples shrivel
ready to be squeezed.

A fleshy pink
creeps like a stocking
up the white hydrangea.

Rivers tugging
hair-like weeds
through mud.

Everywhere I go I find
this grief of living things
that reached their pitch

of ripeness.
Outside my door
red berries swell.

In Childhood

We knock red yellow blue
twilight in and out
of wickets, invisible animals
we need to enter.
The rhododendrons
are out of bounds and all the stars
shine before our colors strike
the final post. Long after

we toss our mallets in the flowerbeds
my drunk uncle plays alone.
His lit Havana hovers
among fireflies.

At The Lake

Skinny Salvador went in first
a diamond flashing on his finger.
Then fat Ken Harris, his voice
already a man's.
Holding my breath
I hardly felt
water or mud, afraid
of brushing their genitals
mysterious as fish
in dark water.
We slipped
with our backs turned
into our clothes
and walked single file
the way we had come.

Father

I

Eighteen years ago you carried me farther
and farther out in the Sound
then threw me
screaming down
and ordered me
to swim back home.

II

A photograph of you, near Marseilles
the sea behind you, the perfect line
where the sky begins.
Your straw hat may blow out
of the picture any minute.
You are looking inland.
I think of how men
after a certain age
seek the sea shore
and avoid the sea.

III

In Proust's hotel the visions
of young girls, petticoats
blown, are gone. Waiters
loiter among potted palms
no one to serve.
The guests have come for cures
the ocean cannot give.

IV

You took me to the casino—red
carpets spreading back to blackness
a rose in your buttonhole *rien*
ne va plus rien the spinning

numbers amused you
a truer gambler
than you think, a poet
afraid of spinning
singulars where there are
no numbers. By midnight

a bouquet
of new francs.

V

You have broken the days
down as far as they will go   Father
show me a sea without
horizons

Mermaid

Waist up, I know.
The rest snakes away:
scales, fins, side-slits.
My thighs inhale the ocean
my wrists and breasts sweat off.
This tail fools the pure fish
the sharks' sweet playmate.

But the manta-ray
the size of a big man's hand
all palm, does his deep dance
I join with only my eyes
the tiniest fish hugging him like jewels.

I'm a woman
simply combing my hair.

Fevers

(In memory, Libby Segal, 1945-1971)

Doctors pinch you
like the last known beast
of a species near extinction.

Your face is a dark moon
in a sky of pillows.
But in your dream
bicycles, motorcycles—
earth is a brilliant wheel
silver spokes flashing
among the stars.

Ward doors
slam and open around you
like wings of giant moths.

Ice is where you long
to lie down,
kick the hot sheets off
and take in glacial air,
your pores clicking
like a million casements.

Your body will leave
a footprint
no one can identify.

You lie awake all night
the only thing that moves
in your body, your eyes
blue tires blowing out
in the dark.

Midas' Daughter

Before my father died outcast
in Eden, choked by gold lapels
he cried for blood and butterflies.
He'd picked his kingdom clean.

I hacked the gilt for dirt
to bury him. Gloss ran too deep.
I put him in his garden
dead beside the dagger lilies.

I inherited his greed,
but not for gold—I'd slept
in solid sheets of it—I fled
the polish of his world for one

where I could work my heart and fingers
to the bone. I asked for it.
Now everything I touch breaks
into tears or flame.

The Nun

I would have died for you
but that was not allowed.

My father left me at the gate
as though I were an infant,
in a sack, without a name.
The sisters cut my hair.
They set my silks on fire.
In these black robes
I was born again.

Mornings in the damp church
stranded with women
as the sun strains
through the glass thighs
of the saints,
I pray for our child
as if she had my life
and for you also
who did what Christ
could not: persuade the judges
of your innocence.

God, Woman, Egg

She never asked to lose her innocence
like this: angels fingering her,
the thrust of doves and roses at her door.
She had been soft and free.
Now the sky strikes her hair
lightning wielded by insomniacs
and her womb becomes the world's:
the fetus in one corner folded
skull and cross bones
like a mouse in a broom closet.

Each day its nails grow longer.
Sperm flows like the Jordan in her dream.
She is God's purse, snapped tight.
She'll shrink as Jesus fattens,
elbows the sac open and slides out,
the spirit, a glue on his tiny thumbs.

The Dump

We cram cans, fish wrappers
into the chevy and drive
through delicate country
to dump it. The late sun's
skinny fingers probe the hills.
Last night we were near
orgy. You danced for the first time
since your senior prom, waving a litre
of champagne. Tornado weather. The wind slaps

blossoms off the trees. We stack the bags
against the cliff and sweat. I feel love
try to sink me
like a sack of kittens. It all goes
in the pit: fat plastic children
snagged on roots. We leave them
ripping open.

Heartland

I

I met you at O'Hare
anxious to be elegant
in my grandmother's
fox fur. You
had no beard and got official
with your bags. I got us
lost in the stockyards.
You yelled
"I'll be a goddamned chauvinist
pig if I want to." I kept my mouth
shut, glared
ahead for smoke.
You switched
the subject to the cosmos.
We were running
out of gas, circling
the heart of the slaughterhouses
and I
was at the wheel.

II

Perfect for a six-foot suicide
the hotel bathtub leaked urine
colored water in the drain.
Rain bit the windows.
We failed in all
positions and blamed
the bed. You blew
your sax and tried to wake the whores
across the way.
Ten flights down
another dozen cars
were lost
and leaning on their horns.

Second Person Singular

I'm learning a language in which you are missing.
It's not a Romance language or a primitive dialect.
It's the kind of language a child invents:
"Don't speak to me until I tell you to."
Its rules are cruel and whimsical.

In my textbook are pictures of the country
where it's spoken. Two-dimensional landscapes:
the first and third. Trains run smoothly, painted white.
Hats are the money makers. Everyone's always
putting them on to go out. Also shoes, made to fit
either foot.

All numbers are odd. When three people gather
etiquette demands the second person remain silent.
Citizens never look each other in the eye
but at a distant spot on the horizon
like swimmers trying to stay in a straight line.
Instead of "Hello" they say "How is she?"
If they want the time they must address the clock.
There's no word for procrastination
or emergency.

After Curfew

All the phones on the block are dead
but if you listen closely you can hear

the microphone in her locket
pick up the wind
in the hairs on his chest.
You can hear the gossip
of another continent:
how he used scissors
on her seams

and how his head hangs
its hollow receiver
between her breasts.

3 A.M.

I leave the light on
so moths keep me company
their lively white bodies
beating through the room
wing, pulse, antenna.

Your eyes are motionless
beneath their lids.
Your body has collapsed beside me
a parachute in the desert
the rip-cord jerked from your spine.

I can barely hear you breathe.
Placing ice on your temples
I kiss you
like an imaginary friend
who will agree to anything.

Snowman

You are hitchhiking from here to Deerfield
your sneakers stained with road salt
your scarf flapping in a car
a hundred miles back.
You know you need a ride before nightfall
fits the ski mask over your face
and you wake luminous in a drift
to find the young girl
shrieking away on her sled.

At dusk as I drive I see you, slouching west
an unlit cigar between your mauve lips
one blue eye, one green
looking as if children had made you.

Raccoon Skeleton At Long Plain Creek

Wading upstream we bump his carcass
with our ankles. He jiggles
like the needle of a compass
scattering silt until he's pure
milk white.
Tin shines through his skull.
He soaks light
right through his sockets.
Water runs its tongue
inside his crevices. We link
fingers, looking down . . .
all the soft things are gone,
bone gets its turn.

Crescent Lake

We drive all morning under the threat
of rain, grey road lined
with nervous fir trees
and your evasive voice:
I'm thinking of leaving.
As though you hadn't thought of it before.

When we pull over you wander off
to meditate.
The water is warm and tugs me
swiftly into the middle
so deep, they say, here
Hell is blue, bottomless,

the lake filled with local stories:
how the drowned turn
into fish;
men dive for a woman's body
only to find her later
hiding under the bed.

If I could terrify you
into loving me I would
go limp and stop
fighting the current
to see you running
towards me just once.

Vashon Island

Autumn came quickly
the entire island
stripped overnight.
Now blue smoke pulses.
Shot pheasant float on the beach.
Carrying the last of your belongings
I feel cool air in pauses
between rain, grit
on the slick macadam
winding down
to the deserted village.

I tie your clothes in bundles
and mail them to you.
Only the town drunk
slouches at the tavern
where we spent every night
murmuring over gin
in the dark booths.
Through the fogged window
I watch trawlers lurch.
Waves make tight fists
and release themselves.

Parting

We hunted pheasant in snow
picked our way through
bittersweet, no birds, only traces
three-pronged tracks
leading to a grove of tiny Christmas trees.
You stopped to pull burrs
from my coat with such care
the way you'd pull
bullets from fowl
telling me not to think so much...

Morgue

Each body is a blue carnation
in a long white box
odorless
as florist-cooled flowers.
They glow
in their greenhouse of flesh
as if lit miles
below the surface
by rose lamps.
Their veins are moths
frozen in that light.
Sound proofed
surrounded by dials
the dead drift
between grief
and earth.

Evening Chores

The garden is growing smaller.
Roses merge with gravel
as the sun withdraws.
This is twilight's ashen grace.
For a moment we face each other
without suspicion, features softened.
Our bodies press against the window
as if to steer this house
across the sinking hills.
Lights flicker on in the distance.
Roofs slide away
like clothes into drawers.
We turn back to our tasks
becoming, whites of eyes,
a shoulder blade
a shirt and a blouse full of wind.

Highway

The fields are filling
not with snow
but ashes
the sky unleashing
its relentless grey
although the moon appears
a solitary bone
on negative.
In the blurred traffic
each driver carries on
long conversations
with himself.
My mind fills
with useless terrors
the fear of darkness
and the fear of giving blood.

Marilyn Zuckerman

Turning Point

For Jane Cooper, Jean Valentine and
the late Tess Wolfson

Contents

Marilyn Zuckerman

Turning Point

Childbirth

First the waters broke
pouring out of me
like the Flood
For two days I walked
with towels draped
around my thighs
Don't worry the doctor said
you won't remember anything

My grandmother had seven children
Five were born dead
I was too small—and O the pain

When it was my turn
I consulted a magician
A man who smiled and said
Don't worry you won't remember a thing

When it was over
I awoke in a large room
There were flowers
My body was empty
my arms black and blue
from elbows to biceps
See the doctor said *you don't remember . . .*

For weeks after
I dreamed of caves—
and a force like the wind
the ocean screamed
water babies tumbled
in their birth sacs
and no matter how I tried
I couldn't remember anything

Dependency

I awake
afraid of my underwear
two dark stains
a musky odor

Afraid of refrigerators
long corridors of white enamel
roaches

I'm afraid of the continent
that separates us
three thousand miles
tires spinning slowly
on an upturned car
cities
under the wheels of your airplane

I'm afraid of the silence
into which water drips
the gurgle of pipes

Lights press against my closed shades
an explosion in the street
rattles the mirror
I'm afraid
of the nick on the door
made by a chisel

I'm afraid of your mother
who is dying
I want to ask you
" . . . is it my fault?"

Ghazal

No one notices I have been smashed against the wall
like a drinking glass;
the lion without teeth encircles me with his paws.

Do not choose me even for love—
The rose in my hair is dying.

You are my purdah screen. I peer at the world
from under the circle of your arms.

A time when life unreels quickly, like a movie or a
carousel.
I can only wait until it stops.

Someone has invented new colors.
I am not armed against them.

Five Days

I

A Quarrel

Tonight I know what I know
Tomorrow I may forget
Tonight I am a gambler
five bullets in the chamber
Tonight I walk a tightrope
Tonight
I roll dice
The stakes are
All or nothing

II

Anger

When the red eye of love goes off like a hand grenade
don't blame me if your hair is singed.

Two cats, one grey and one black, chase each other
up and down the stairs like love and anger.

If you loved me you would be here
to eat the bread I baked because I was alone.

This Russian music is sad as a drunken Cossack.
Cello notes in a minor key gnaw a hole in the air.

III

Fever

I know
Sometimes I am a terrible disease
Something that never goes away

always
burning
burning
like a fever of a hundred and six
Anger and love flame up
Some mornings
the ashes
lie in heaps
on the bed
the floor
in corners of the room

IV
Red
Remember how it was with us that night love
Outside
fire engines slammed
somebody shouted
Red
He was wearing red

Is happiness too terrible to bear
Does it steal like a thief
across rooftops—
drop the swag
escaping up Madison Avenue
—lost in the night

V

What If

What if I love you
as I love the grass
for greenness
and cool moist roots
As I love evening
that wraps around me like water
I absorb the earth
                    with
                              fingertips
                              mouth
                              the roots of my hair
                              and the pores of my skin
I move across time like a tortoise
I have nothing else to give

Before Completion

Before Completion
is the exhaustion of
the masculine
—I Ching

I

Elegy for my Father

Take him out of the earth
let the sea clean his bones.
Only water will wash away
The stain of mustard gas breathed in the woods
of Chateau Thierry
of too poor to go to law school
The dust of drab offices
Shreds of green money
The smoke of burning subways
The fear of great mountains
fear of mountain passes
fear of long journeys
of deep oceans
of handshakes
fear of breaking wind
fear of the dark road
that disappears into the forest
The taste of rich food
The closeness of warm rooms dresses thick perfumes
and too many eyelashes
The stink of medicine
The drip of stomach acid
and mucoid fibers growing wider
at the bottom of his lungs

Take him to sit under a tree
and hear the swarming of bees
Let him become a tree—
His hair matted with birds
His tongue tasting the tartness of roses
And his voice mingling with the wind

Let him grow wings like the heron
Carry wild fennel in his beak
Rosemary and cherry blossoms
tangled in his feathers
Let the Alps be in his eyes
Let his shadow darken the Rhine
the Amazon and the Nile
Let him shake off the crumbs of ancient books
caked in the corners of his eyelids
        the oil of smoking lamps
        and ear-locked mumblings.
Let him become a lion
Pawing the spongy, fragrant earth
Springing from rock to rock like a conqueror
A deer sniffing the wind at a late-afternoon pool.
Let him rise like a redwood in the forest
Let him leap through a hole in the sky

II

Father

You keep coming back
laughing into my dreams
with large, crooked teeth.
No grave can keep
your terrible angers
In the early morning mirror—
in the twilight window
I see your death's-head grin

Dialectic

I
Husband
Dear familiar—
my good bread
true as grass
I mean to say
you are a homely thing
an earthy thing
—better than mother's milk
now you've
thrown away your
three button suit
riding crop
and stopwatch
Lighter than any youth
tenderness
has a masculine gender
To husband
is to make things grow

II
Tower/Yale University
No matter how I stare at it
It will not melt
   will not go away
become something else
something less present
less awful, less dense
It will not smile
   produce metaphors
   become playful
or give up its power
It only grows larger—

Dissolution

I
Our old life strewn at the curb
picked over by the doormen next door
They wear our clothes
read our mail
laugh at our diaries

II
The form breaks up
becomes a cloud
contracts
leaves a void

III
I sit up nights
listening for my life
Waiting to go down to it
to the slowness
and stillness
where everything happens

IV
Learning
to let it move
slow and cool
like water

V
My life
on the
doorstep
drying in the sun
like a starfish

VI
Moving among familiar things in strange rooms
waiting for sharp edges to go away.
Learning the pattern a leaf makes against the wall
or a window that casts a watery reflection
Searching for an old diagram of the city
listening for its rivers
                    its crickets
Hearing the grass rustle at night,
            the chestnuts falling like footsteps
            the small talk of the streets
Waiting
to turn at last
above the relief map of the city

Turning Point

I
Late Afternoon
Shadows crisscross over me
binding me to this garden
where a raccoon is draped
asleep
in the crotch of a tree
like an old fur scarf
and only I am homesick
wanting the sea

II
Evening
I was taught from childhood
to be home before dark
Now the dark
has become
    a mother—
    a husband
fetching
    a lantern
    a cloak
    and my shoes
Twilight calls me home
I slip down rushing streets
like a sailor in a foreign port
afraid his ship will sail without him

Hurry
Hurry

III
Sexual Politics
We lived behind windows made by others
Now a thin bullet breaks them
Who cares for dogwood trees
spring's green, seeping pollen
drifting over everything
burying our lives in a garden
when what we want
isn't born yet

Sisters
I want to cradle
your life in my arms
pour unguents
mutter witch charms
over your wounds
I want the moon.
The sun has long been theirs

IV
What Collette Knew
After middle age
a sensous woman
is worse than
a dirty old man
Suppose that all I wanted
was to touch your hair
your shoulders
Suppose you look at me
as though I were
ten years younger

V

The Third Wish

The sun to stand still
The sea at my side
and you
looking at me
looking at me
looking at me

At the Party

While
shadows
of dancers
slide across the wall
we smoke
together
and listen
Feeling
without
touching—
a magnetic field
grows around us
These words are
not my body
but enclose us in a tight space.
Your words will not
become a dark street
—yet have the power
of an old story
appalling,
almost irresistible
Like witnesses
to a
fatal accident
there is nothing
to which we can agree
Brother
outrage is numbed by your eyes
and the way you try
to keep them
from leaving mine
We speak together
We crack open a very small place

Variations

I
Why I Wanted You
It was accidental
Like being lifted
by the spout of a whale
eight feet of water
and me turning and sputtering
under it

II
Déjà-vu
All the flotsam
of the older affair
floating—
film props
waiting to be used
for this one
The same brown eyes
expecting answers
the same sexual glow
bursting
like a blue flame
around your hands
Now that it's over
now that it never happened
I know
how little
it all had to do with you

If Life Hands You A Lemon; Make Lemonade

If you are motherless; nurture yourself
—an orphan; make do
If you are lost; go somewhere
                    find yourself in the silence
If you have no dreams; invent them

If friends leave
lovers abandon
and the throbbing
in you
waits for the underbeat
take that fierce energy
hammer it into a poem
—and give it away

Pond: Vermont

The pond is the eye of the landscape.

—Thoreau

A hot day
a few wrong turns
a green pond
blooming with hyacinths
the white coolness
of naked bodies
Two young men
meditating
spines growing like trees
along the bank
whole families
bare as flowers
fathers
with huge red uncircumcised cocks
children
slipping like dolphins
between each other's legs
mothers
long wet hair
curved in an S
around their breasts
Susan and I
took off our clothes
and the cool water came in us
two farmers—
one middle-aged
the other younger
penises limp
as though they never took off
their overalls before
—looked at us

but we were peeled branches
white logs stripped of bark
Later
drying on a towel
I felt
the sun
on my nipples
for the first time,
my legs part
to let heat
in

Atlantic Fog

Land's end
so deep into the Atlantic
seabreath
hangs on all things
on trees like Spanish moss
in a cloud
smoky as Saturn's rings
And there are days and nights like this
when fog
stands before the fire
wetting tiles—
a drowned sailor
home at last
It dims lights
hurls swallows and seagulls
against windowpanes
curls a noose around trees
and strangles the sun
Net poles in the abandoned weir
bulk out of the mist—
and the island opposite
drifts like an iceberg
out to sea

For Marcia

Part boy—
all woman,
you are transparent
I can see
blood vessels
branch
to your fingertips
You stalk your rooms like a terrorist—
the most efficient nurse
in all the Kildare movies
You trace the profiles
of your children
in the dust of the coffee table
and smell the perfume of their hair

Angel of Anger
You face the loneliness you dread
with a sword
Mermaid
exchanging fins and tail
for the right to walk
on knives

Mediterranean Women

after Akmadulina-

Black Indian summer flies
thick as dead leaves around my bed
The cat plays with one
still buzzing.
The kitchen is dark—
summer of '68
the Mediterranean
wind-twisted pines
fine black brows
skin, amber as honey
the Persian poet, Hafiz
Sufi mystics
late night talks
and the music and perfume
of Mediterranean women

They walk delicately on long legs
—camels crossing the desert.
Carillon laughter
soul deep, black eyes
and names like
Sora, Nicol, Dunya, Elena

Sora reads a book by candlelight
Nicol walks the beach alone
Elena laughs with a child
Nina is afraid of men
Dunya doesn't trust them

Dunya,
the streets of Persepolis
a door
set low into lintel posts—
she knocks
and hears the cries
of baby girls
left to die
in the Mesopotamian desert
the high wail
of Berber women

Nicol hates the cold
shivers in the wind
blooms in spring
is gay in summer
moves with American strides
through Northern cities
Mediterranean languor
heavy on her eyelids

Summer—
jasmine and honeysuckle
whisper their names
Nicol, Dunya, Sora, Elena.

Now

Now we begin to live
the life
we wrestled for with angels

dawn
silent rooms
a single lamp
circling lined paper
over which a pen skids

Don't look back
don't keep journals
ledgers of losses
desires for childhood
or peace

Now we begin to learn
what farmers have always known—
grit between the teeth
the hard, obdurate gaze of the sky
landscape stretching to the horizon